# How To Eliminate The Pain of I.T.

THE ULTIMATE GUIDE TO TECHNOLOGY FOR HEALTH
CARE PRACTICE OWNERS AND ADMINISTRATORS

John Verbrugge

Nonlinear Tech, Grand Rapids, Michigan

John Verbrugge/Nonlinear Tech
P.O. Box 6094
Grand Rapids, MI 49506
www.nonlineartek.com

Book Layout © 2015 BookDesignTemplates.com
Photography © 2016 VeronicaKirinPhotography.com

How to Eliminate the Pain of IT/ John Verbrugge. -- 1st ed.
ISBN 978-0-9978837-0-1

# CONTENTS

# CHAPTER ONE

# Your Health Guide For Technology

In the year 2000, I was asked by a friend to help out a small organization set up their first local area network. It consisted of three Macintosh computers and a printer. That was the start of my tech support business. Over the next decade and a half, I grew that business into a successful technology consulting firm, and improved my knowledge, my processes and my philosophy of I.T., year over year.

Now, 16 years later, I decided to use my hard-earned knowledge of Information Technology (I.T.) and my understanding of the challenges of running a Small Business to create a health guide to help the Small Business owner make good technology choices.

Small Businesses and clinics really want to do the right thing with technology, but there's often no one to turn to for an objective opinion. How do you choose a good I.T. firm? How do you know you're getting good advice? How do you know if your I.T. person is doing a good job?

I want to share with you the "healthy I.T. choices" that can make your life easier and ensure your I.T. decisions are the right ones for your business.

Consider this book a Health Guide for Technology that will help eliminate the pain of I.T.

# The Four P's of Wellness-Focused I.T. Support

## Protection, Price, Presence and Practices

When I talk to Small Business owners, I often ask what their greatest fears are regarding their technology. I get a variety of answers, but they boil down to a few major fears: the fear of whether the business is protected from a disaster; the fear of price—of being over-charged; the fear of non-presence—of not getting support when you need it; and the fear of practice—of not having up to date technology or advice.

### Protection—Are We Protected From Disaster?

Sometimes, the questions are these: We paid for data protection. Why did we lose our data? Why did we get hacked? Or why did we get a bad virus infection?

Technology is supposed to make things easier. Why go to digital X-rays instead of paper? Why use a database instead of paper files and folders? Digital X-rays and databases are more convenient, can do many more things, and ultimately save time. But digital data comes with some risks, including the risk of loss or corruption.

When a Small Business moves into using technology to run the business, the transfer usually happens slowly over time. I'm old enough that I remember organizations that used typewriters for some of their forms and other specialized documents. I helped them convert those documents and forms to PDFs. I helped those same organizations purchase servers, storage area networks, enterprise backups and so on.

Over time, those organizations have come to do everything digitally.

Now imagine if, instead of working with an I.T. professional, they decided to try to take care of all technical issues themselves, but investing no money into infrastructure or data recovery. Imagine if the server crashed, and they had no backups–all would be lost. The entire organization would come to a standstill.

There are businesses and practices in that situation, and that's very scary.

But it goes deeper. What if, all along, the Small Business had been working with an I.T. company who wasn't doing a good job? They had *trusted someone else* to protect their data, but they *still* lost everything.

That's very, *very* scary.

This fear takes many forms, but it comes back to the fact that they have paid an I.T. person or organization who might not be doing the right things to take care of their digital assets. Business owners don't know if they are protected or not.

Do you know if your business is well protected?

**FREE I.T. REVIEW:** Get the peace of mind and assurance which only a *properly conducted* I.T. review can provide. Ensure you're being well taken care of—*even (or perhaps especially) if you already have an I.T. support service.*
www.nonlineartek.com/IT-review

## Price—Are We Over-Paying For the Support We Are Getting?

This is a bit less scary than not knowing if you're getting good support, but it still makes business owners nervous. What's the right amount to pay for I.T.? That could be a book all its own, but I can summarize it this way: Companies that are willing to be proactive about their I.T. investment and make good support a priority do well.

How do you know if you're paying too much for support? Or if you've hired the lowest price cheapskate I.T. firm? Get a second opinion.

I've been the "second opinion" I.T. expert for years, and am happy to give business owners a second pair of eyes to look over their I.T. expenses.

Pricing in I.T. can vary considerably, and price isn't necessarily tied to quality. Some underperforming firms charge high prices, and some excellent firms are very economical.

Without a second opinion, you may go on for years overpaying for your I.T. service. If you don't get a second opinion, you'll never know, right?

**FIND OUT IF YOU'RE OVERPAYING!** Request a second opinion review of your current I.T. services and charges. Find out approximately where you rank compared to local and national cost averages. www.nonlineartek.com/second-opinion

## Presence—Where Was Our I.T. Guy When We Needed Help?

This is one of the most frustrating things a company can go through regarding I.T.

I have a friend who owns a restaurant. When his email got hacked, all he had to do was change the password for his email account, and the immediate problem would be resolved. It took him 8 hours to *find* his I.T. person. Wow. If you can't find your I.T. person, you need to spend some time to find a *new* I.T. person. In the cases of an I.T. emergency, being able to reach your I.T. team and know that the problem is being worked on is critical.

When your clinic or Small Business is having a technical emergency, every second counts. If a virus or other malware is corrupting your data, an hour can make the difference between saving your data and having everything destroyed. Getting immediate support for a server crash on a day booked full or patients can make the difference between making a profit and showing a loss, between helping patients and being unable to help them. If some of those patients are new patients, they might never come back.

During a technical emergency, you *must* be able to reach your I.T. service quickly.

Most good I.T. firms have three or four ways to get a hold of them—phone, email, web portal, and some will accept text messages as well.

Let me also say that it's the I.T. company's responsibility to clearly communicate the best way to get a hold of them. In my company, Nonlinear Tech, we have a company phone number, a support email address and a support web portal. We have clear specific procedures for handling I.T. emergencies, and our customers are clear on what to do when they need to reach us immediately. The phone, email and web methods all create a support ticket automatically. That way, either I or one of my technicians can act on it immediately.

As a business or clinic owner, you need to have clearly spelled out methods from your I.T. firm for contacting them. When you do, you should also receive some feedback from the I.T. firm that the message was received, and that they will attend to an emergency quickly.

> **I.T. CONTINGENCY COMPANY:** What happens if your primary I.T. service is unreachable or unable to solve a particular problem? You can't afford the downtime. That's why smart businesses keep an I.T. Firm on contingency...a tier-2 I.T. service that can step in to help if the worst happens. See how to keep your existing I.T. firm AND have a professional I.T. service ready to step in to help at a moment's notice.
> www.nonlineartek.com/backup-IT

## Practices—Why Was Our I.T Guy Not Using Best Practices?

This is another fear of business owners. Sometimes they ask "How are you keeping up with all of the changes in I.T.?" In other words, "How do I know you are going to give me current advice?"

Fair question. I know several I.T. guys who stopped learning 8 years ago, and are just keeping everything running the old way. In my case, we use a variety of methods to stay current. We go to I.T. conferences, attend I.T. meetups, we attend product and service presentations, we meet in tech mastermind groups, and spend time every month learning something new.

What happens if your I.T. person isn't keeping up? You might be paying more for some services than you should, because many services have come down in price. You might be receiving old technology or be using systems that are out of date.

Even worse, you might be using software (such as antivirus or antimalware software) that is worthless and ineffective. Imagine paying your I.T. firm for antivirus software that is ranked 9$^{th}$ or 10$^{th}$ on the list of best software. The consequences of having poor antivirus software are devastating. One malware infection can wipe out your whole hard drive! It's inexcusable that some of these I.T. firms haven't spent the time to find the best solutions; their negligence puts your business at risk.

The I.T. world moves fast. An I.T. firm needs to stay current. The most valuable thing we do is dispense technology advice to our clients.

A good next step would be to download the best practices checklist to rate how well your I.T. firm is staying up on current technology, and then ask your current I.T. firm these questions.

**IS YOUR FIRM FOLLOWING *TODAY'S* BEST PRACTICES?** Compare your I.T. firm's work with this continually updated I.T. Best Practices Checklist!
www.nonlineartek.com/best-practices-checklist

The Bottom Line Is Trust.

The business owner needs to be able to trust their I.T. firm, the same way they trust their CPA. In fact, the I.T. service might be even more important than the CPA. The CPA helps you keep track of the money, but the I.T. firm keeps the "money machine" running. It doesn't get more crucial than that!

As a clinic or Small Business owner, surround yourself with high-quality, recommended vendors and consultants, and understand that the money spent on good I.T. is an investment in the future of your business.

# CHAPTER TWO

# What Is I.T. Work, Anyway?

Let's take a quick side-trip to talk about what I.T. work really is. We can think of it in three parts: Creating your data, Accessing your data, and Protecting your data.

## Creating Your Data

Creating your data involves getting the right software, learning how to use it, making sure everyone in the organization knows how to efficiently use the standard office software and the specialized software for their industry.

This is where most Small Businesses do well—they know they need Microsoft Office, an email client, the accounting software, and perhaps specialized software for their industry. Employees are hired based on their existing skills with the software. Everyone could use more training on something, but in general, the creating of data is the easy part for the business.

## Accessing Your Data

Accessing your data includes the internal network, storage of files, the server, and how your workstations are connected. It involves WiFi, using smartphones and tablets, VPNs and cloud storage, and the internet provider for the organization.

Getting to your data is the part of I.T. that is invisible to most companies; that is, when it's working, they don't see it, and when it's not working, it's a Very Big Deal.

The I.T. firm helps by making sure connectivity is there, by ensuring that all of the parts are working together, and by solving connection problems quickly.

## Protecting Your Data

Protecting your data means making sure backups are done right, stored correctly and accessibly, and that the right data is being backed up. It also involves network security, firewalls, intrusion detection and preventing malware and viruses from infecting systems.

The I.T. firm is least visible but most important here. Backups are never in the forefront of a company owner's mind, but they are the most critical thing that an I.T. firm can do for their customer. Likewise with network security.

## Why is understanding this concept important?

The concept of creating, accessing and protecting the data is important because business owners typically have their priorities in exactly the reverse order. To a Small Business owner, the most important thing is running the business—creating products, billing for their services, and creating data. The Small Business owner usually doesn't give a second thought to how well the backups are done (until it's too late, of course).

A top tier I.T. firm will focus their efforts toward network security and protecting the data. This focus ensures that your Small Business is protected, and even saved from ruin in worst cases.

Is your I.T. firm taking care of this critically important part of I.T.? Are you absolutely, completely, 100% sure that your I.T. firm has network security and data preservation taken care of? How do you know? Have you asked them? Do you know the right questions to ask?

**DATA CREATION, ACCESS, & PROTECTION WORKSHEET:** Is your business adequately protected in all areas of data creation, data access and data protection?  Download the worksheet and see.

www.nonlineartek.com/create-access-protect

# CHAPTER THREE

# Take the I.T. Wellness Test: Eight Critical Checkpoints for Your Practice

Preventative maintenance is a well-understood concept. The cars that have the least trouble have had frequent oil changes and inspections. People who have few dental problems have had regular checkups all their lives. Similarly, businesses with the least computer trouble have had I.T. preventative maintenance done by professionals for years.

The old philosophy of "Call me when it breaks, and I'll fix it" is asking for trouble. In the I.T. world, we call this "Break/Fix". Many I.T. firms still operate this way, although the successful firms employ many preventative maintenance tools and techniques.

The industry buzzword of "Managed Services" incorporates the idea of preventative maintenance into a flat fee pricing model. In other words, Managed Services stresses the importance of preventative maintenance of the computers and networks – monitoring the systems, making sure updates and patches are applied, doing work remotely, and so on. We'll explore this topic in detail shortly. The other piece of Managed Services is a flat-fee billing arrangement per workstation and per server, so that that customer pays the same amount every month. It's similar to the idea of insurance.

As we'll see in a minute, the ideas of preventative maintenance and flat fee pricing are not the same thing. Good I.T. firms can offer excellent proactive support, using any one of several billing arrangements.

My philosophy for Information Technology is called "I.T. Wellness". In the same way that you go in for dental checkups and take vitamins to keep you healthy, you do proactive work on the technology systems to prevent failure and outages. That's the wellness model.

Below are eight checkpoints that you can use to confirm that your I.T. firm is serving you well.

## Wellness-focused I.T. Support is Insanely Organized

One aspect of I.T. Wellness is being "insanely organized". By that I mean we keep very good track of customer I.T. requests, the status of projects and tasks, as well as customer equipment, and the history of work done for a client.

I know I.T. professionals who keep "everything in their heads." That is, nothing is documented, there is no system, and client information is haphazardly scattered around, and often out of date.

Let's take two hypothetical I.T. firms and compare them:

Carl's Crazy Computers: Carl keeps his list of customer master passwords in his head. His clients call him on his cell phone, or text him, or email him, and he just responds when a request comes in. He's responsive, and gets back with them right away, and keeps their systems running. He's a good technician, and keeps track of billable hours and work done on a paper notepad, which then gets copied to an invoice later on. Carl has some Word documents with other passwords and server information, and he gives that information over to vendors and contractors when needed, but it's often out-of-date information. If

asked, he can usually remember the right information.  Carl creates and sends out invoices monthly,
but if a client asks for more detail on a particular
item, Carl can't usually find the information, because
it's not in his notes, or he just doesn't remember the
details.

Let's compare that to Tom's Top Tier Tech.

Tom uses a ticketing system for incoming support
requests.  Clients can email or use the phone, but either way, the incoming request gets entered into the
ticketing system.  Clients can then log into the ticketing system portal to view all of their tickets, and
check on the status of any ticket they have entered.
Tom keeps password in a password vault program.
He can delegate who has access to passwords, and
can track who accessed a password at any time.  Tom
creates standard operating procedures for anything
he needs to do more than a couple times.  Any employee or contractor who helps him must follow the
standard procedure.  His employees and contractors
are consistent and efficient.  Client information is
stored in a knowledgebase that he, his employees and
his clients have access to.  Tom's ticketing system
generates invoices from the tickets; any time a ticket
gets worked on, the hours are logged for that ticket,
and all of that detail is available to the customer both
before and after the invoice is generated.  Tom is organized, efficient and has a system that is nearly
automated.

Both Carl and Tom are running profitable I.T.
businesses, and Carl might actually be *more* profitable
than Tom in the short term, because he doesn't have
the expense of the ticketing system or spend time on
documentation.  But if anything were to happen to
Carl, his clients would suffer a great deal.  Tom's I.T.
business is set up for success—he produces consistent

results, and can continue keeping his clients running smoothly, even if he needed to take extended time off.

I.T. Wellness uses the systems and tools that Tom does—insanely organized and efficient.

The consequences for the Small Business owner of working with a disorganized I.T. firm are that you will become frustrated with the technology in your clinic. Problems won't be resolved on a timely basis, you'll have to call twice to get something fixed, the technician will have lost critical passwords, and simple issues will just take forever to get resolved. Frustration will build over time, and you, the business owner, will feel like something's always broken, that problem-free day is unattainable. It will affect your finances too—spending extra time on I.T. and having employees hampered by technical issues will make your business less efficient and less profitable over all.

Get organized, and get an organized I.T. firm on your side.

### Wellness-focused I.T. Support Means Doing Things Right the First Time.

The one thing that wastes the most time and resources in any company is having to re-do work that was already done but done poorly.

So I.T. Wellness strives to get things done right the first time. Using a combination of documented procedures and real-time knowledge of a client's network and computer systems, the I.T. Wellness system gets the issues resolved and the projects done right the first time.

The consequences of careless I.T. work is that you'll pay for it twice. Even if you're in managed services, you'll pay by having extra down time. It's not worth it to have a careless I.T. firm caring for your systems.

## Wellness-focused I.T. Support Means "The Accountant, Not the Plumber."

This isn't meant to disrespect the profession of plumbers, but one analogy I love to use for I.T. is the "accountant versus plumber."

When you call the plumber, you call when there's a problem, you need him to show up right away, fix the problem and then leave. You typically don't ask for advice, and you don't keep him around to advise on the direction of the company.

The accountant, on the other hand, is someone you meet with regularly. She has the pulse of your company, you trust her to advise you to make good decisions, and you involve her in all of your major company decisions. She is one of your trusted advisers and is crucial to your company's success.

Your I.T. Wellness firm is also a trusted adviser. I.T. Wellness is part of your planning process, and your I.T. professional is there to give you advice, direction and ideas to help your company succeed. Proactive work is part of the planning process *before* a project gets underway, and continues on the entire schedule of the project. I.T. Wellness means planning ahead.

The consequence of having a "computer plumber" I.T. firm is that you never develop a long-term strategic relationship with them. You're missing out on the greatest value of a good I.T. firm. The most important thing the I.T. firm does is dispense good advice and direction about very complex topics. If your I.T. firm acts like a plumber, you're losing out. A strategic I.T. partner is someone that can actually help your business grow.

QUARTERLY REVIEW CHEAT SHEET: See how to improve your security, speed of business, and bottom line with this FREE Quarterly Review Planning worksheet. Download here. www.nonlineartek.com/quarterly-review-worksheet

## Wellness-focused I.T. Support Means Understanding Trade-offs.

You've probably heard the old adage "Good. Fast. Cheap. Pick any two". It basically means that there are trade-offs in any solution you choose. If you want a top-quality solution, and you need it now, it won't be cheap. If you want a cheap solution, and you want a really good one, it won't be ready this week.

*Good* solutions take time and cost money.

I.T. Wellness seeks to help the Small Business owner understand the trade-offs, and make educated decisions.

Once when I was working on a new building project with a client, the client insisted on using the cheapest wiring contractor; they whipped through the wiring project quickly, but after they left, we found dozens of wiring problems, from mis-wired jacks, to unlabeled wires to entirely missing cables. Unfortunately, this client wasn't willing to listen to good advice, and we spent a lot of money afterward fixing the problems with the wiring.

If your I.T. firm leans too far in one of these directions—good, fast, cheap—then the other parts of the equation suffer. Too cheap means quality suffers. Too "good" means you over-pay or pay for what you don't need. And so on.

The bottom line is this—know that good solutions are either costly, or take longer to implement, and that by understanding the trade-offs, you can make the best decision.

## Wellness-focused I.T. Support Means Automating What Can Be Automated.

As we saw in the comparison of two I.T. firms, automation is part of what makes a good I.T. Wellness firm. Many processes in I.T. can be automated; automation makes things cheaper, more reliable and easier to maintain.

An example is an automated backup reporting solution. The I.T. Wellness system has a birds-eye view of all backups for all clients. We can see at a glance which backups succeeded, which ones had warnings, and which ones failed. From that birds-eye view, we get instant feedback on how backups are doing over all; it's automatic, and requires no manual work to retrieve that information.

Another example of an automatic system is a support ticket system that is transparent to the client. When a customer calls and requests help on something, it goes into the ticket system. The ticket remains open until the problem is solved. Every time a technician works on that ticket, he or she logs the time spent on it. At any time, the client can log in and see where time is being spent, and the status of any issue. The reason it works is that it is automated. No manual copying and pasting of spreadsheet data, and no manual entry of hours or work done.

The I.T. Wellness system uses automation as much as possible, so that time can be spent on more productive activities.

What if your I.T. firm doesn't automate? The consequences of having an I.T. firm that relies on manual processes is that things get missed, and stuff doesn't get done. It's similar to the organized versus disorganized I.T. firm. If the I.T. firm is organized, they don't lose things. If they're automating systems, they don't lose things. If they're disorganized and doing everything manually, they'll miss important details and won't realize that a backup has failed. That missed backup might have been the only one that had your digital X-rays for the week. Is it worth it to keep using a company that is clearly in chaos?

## Wellness-focused I.T. Support Means Being the Customer's Stable Partner.

The I.T. Wellness system works for the long-term satisfaction of our customers. In my own business, I have had some of my clients for fifteen years. In some of those organizations, nearly the entire office staff has come and gone, but the I.T. service has remained constant. We have become the one stable partner for those businesses; we are the valuable resource for the organization's knowledge, its history, and its business processes. Well over half my firm's business comes from clients that have been with me more than 10 years.

Why does this matter? It shows that I.T. Wellness means we take good care of our clients, and they have benefitted from it. When you choose an Information Technology partner, find out how long they have been in business, and find out how long a typical client relationship lasts. A firm with many long-term clients is likely practicing Wellness I.T.

## Wellness-focused I.T. Support Means Having Personal Contact.

While I.T. Wellness seeks to automate things and create efficient systems for I.T., we also know that the best support we can offer our clients is personal. That is, we know every customer, we meet with each one regularly, and they know each of our technicians.

Personal support eliminates frustration and makes things more efficient for the business owner. Efficiency means less wasted time and effort.

Early on in my business, one of my biggest clients switched to a larger I.T. firm because they needed 24x7 coverage (or, they thought they did, but that's a long story). I visited them a few months later, and I asked how things were going. The office manager was really frustrated with this larger firm because each time they sent over a technician, he was someone different who didn't know the history of their systems. She would have to explain the problem again to the technician, and the technician would invariably try something that had been tried before, and the problem wouldn't get solved, and the situation would repeat itself.

I.T. Wellness also means Personal I.T. We don't treat our customers like a number; we spend the time to build the relationship.


## What I.T. Wellness Is and Isn't

I'm a big fan of taking care of small tech problems before they become big problems, and of preventing problems from occurring in the first place. I believe that is what sets a good I.T. firm apart from a mediocre one.

One of the incorrect perceptions I have heard recently is that if an I.T. firm isn't doing managed services, then that firm must be doing "reactive sup-support".

Some unscrupulous I.T. make a claim such as "We come in and take care of all the issues on your computers at the beginning, so that there won't be any problems going forward."

Really? Wow, that's amazing support. Somehow, they can anticipate and fix every problem before it occurs.

True I.T Wellness support sets up systems that ensure that issues are taken care of early.

Proactive I.T Wellness support includes the following concepts and activities:

- Perform **remote** monitoring of all servers and workstations

- Do **automated** updates and patching of Windows and 3rd party software

- Conduct **regular maintenance** of servers and workstations

- Keep a view of the **whole** network and computer infrastructure

- Make **planning** a part of the process: that is, having meetings and planning sessions for the future of the whole system

As mentioned before, the older "Break/Fix" model for I.T. support where the I.T. firm waits for a support call, and then goes out to fix the problem doesn't really work anymore. I do get calls from time to time from companies looking for someone to come in and do a quick fix for something that isn't working. If I take these calls at all, it's at a much higher rate and I use the situation as an opportunity to present them our I.T. Wellness support plan.

If a prospect rejects the idea of working with us on a proactive support basis, we don't do business together.

Getting back to that idea of fixing everything before the problems occur, I have some thoughts on that.

There's a difference between an I.T. firm whose *philosophy* is reactive in nature and a firm who is proactive in nature, but responds to customer requests. Every I.T. firm has to "react" or respond to customer requests that come in. It doesn't matter what kind of business model the I.T. firm has set up; there are requests that come in, and those requests need to be responded to.

The goal of Nonlinear Tech and any firm that's trying to stay ahead of issues is that we set up proactive support systems to minimize technical issues that may arise down the road.

Secondly, I find that some managed services I.T. firms confuse the idea of proactive support with managed services. Essentially, "managed services" is a billing arrangement. That is, the managed services firm bills the client at a flat rate per PC or server, and agrees to maintain the computers and network on that flat fee. They may be doing proactive support (I hope so anyway), or they may be very reactive.

Managed Services is a "bet" that the I.T. service firm makes with the client in which the I.T. firm thinks it can do less work for more money by charging the client a monthly flat fee instead of another billing arrangement. The client accepts the bet, because they often believe they can get more work done for less money than with an hourly arrangement. Just know that Managed Services isn't always the best choice for a Small Business, and that, regardless of what the Managed Services firm tells you, it doesn't always save money. For some Small Businesses, Managed Services saves money; for other businesses, Managed Services is more expensive.

For all businesses, having an I.T. Wellness mindset is the *right* way to care for your infrastructure.

For my clients, we offer the proactive support that I call I.T. Wellness. In fact, we don't do any other kind. We're proactively taking good care of them.

# How To Keep Your Technology Safe and Operating At Peak Efficiency

In this section, I would like to expand on the ideas of Wellness-focused I.T. and provide some criteria for evaluating an I.T. firm. If your I.T. firm is a good one, they will keep your technical systems operating efficiently and your data safe.

There's a wide range of quality in technical support firms. Some I.T. firms are right on top of things—organized and knowledgeable, using best practices and excellent customer service. Other firms are, well, not so good. When I talk with business owners of companies who are not (yet) my clients, I'm astounded at what some of them put up with—no-shows at appointments; technicians that are impossible to get a hold of; invoices that are 6 months late or void of any detail; disorganized and lost information. It makes me realize how fortunate my customers are (if I do say so myself). ·

I want you to know a few of the marks of a top notch I.T. firm. Read through the items here, download the checklist, and see how your tech firm compares.

### Does Your I.T. Firm Catch Small Problems Early?

Whether they are catching small backup errors before they become a backup failure, or whether they are monitoring your computers and can fix a small problem before it causes a system crash, it should be clear that staying *ahead* of problems in I.T. saves money.

The most expensive projects we do are when we have to clean up an organization's I.T. that has been neglected or managed haphazardly for a number of years. The amount of time and money spent on fixing and cleaning up neglected infrastructure is twice what it would have cost the organization to stay on top of things and maintain their systems.

If your business is behind in I.T. maintenance, request that your I.T. firm spend some time to fix the small problems this year, so that going forward, that looming disaster won't happen. What kind of looming disaster? If your server hard drive is sending out "failing drive" errors, but you are ignoring them, the drive may fail and you will lose everything on it—patient records, accounting data, images, important documents—everything. Or if your router is old, it could fail at any time, leaving you without internet access. For some clinics, that means no access to any patient data for at least a whole day.

Most I.T. firms offer free or inexpensive system audits, and give you a list of the issues to fix. You can start there and know that you have saved your company a great deal in future I.T. costs.

### Does Your I.T. Firm Meet With You About Strategy Frequently?

I've found that companies that have the least trouble with servers, backups, viruses and technology in general are the companies that meet with their I.T. person frequently.

It's like getting oil changes and going to your dental appointments when you are supposed to; if you want your technology to work for you, you need to meet with your I.T. person once a month. Maybe more.

At these meetings, you can get updates on your company's security and other open issues. The more often you meet, the better your I.T. infrastructure will be maintained.

I've had a couple "clients" (who weren't clients for long) that wanted our relationship to be strictly call-as-needed. Of course, they would only call when there was a big problem, and in many cases it was too late to do anything about it, or it was a very expensive problem to solve. It was clear that they didn't really care about their infrastructure, and that made it hard for me to help them.

If you are involved in your company's well-being and you *want* to keep things running well, you're in a good position to have I.T. well maintained. We call this "taking an active role in the care of your system." On the other hand, if you want to ignore the network and servers until there is a problem, then you will get a system that slowly degrades until there is an emergency.

Ask yourself if you want your business to succeed in the long run. If you do, make I.T. an integral part of your clinic's future plans. Don't leave it to chance.

### Has Your I.T. Firm Performed a Security Audit For You?

Do you know where your security holes are? Wouldn't you like to know (and then fix) any huge vulnerabilities that might be in your business?

Just remember that cleaning up after a security breach is at least five times more expensive than preventing it in the first place.

A security audit might cost a couple hundred dollars, and fixing a few security issues might cost about the same. But cleaning up after a virus or intrusion can cost thousands of dollars. A server infected with a "crypto" virus may take weeks to clean up, including recovering files from backups, checking each folder, comparing documents, testing documents, and more.  During that time, the company is in complete chaos—critical business files are unavailable, employees sit idle while their data is recovered, and everyone spends an enormous amount of time trying to remember or figure out which files are the most recent, which files are important, and which files will need to be re-created from scratch.  Lost time and productivity costs the business many thousands of dollars, in addition to the costs of I.T. services.

A simple security audit can prevent the possibility of an infection.  It's an easy decision to make that will save a great deal of money, and also make your network much more secure.

**FREE SECURITY AUDIT WORKSHEET:** Find out if your company is vulnerable to the latest threats by downloading the FREE Security Audit Worksheet.
www.nonlineartek.com/security-audit

## Does Your I.T. Firm Monitor Your Systems Remotely?

Your I.T. firm *is* monitoring your servers and workstations, right?  Only a few years ago, this was clunky and costly, but now there are so many good tools for computer monitoring.  Now it's a no-brainer.  This is a core concept of the Wellness I.T. philosophy.

Why monitor the computers?  Your I.T. firm can know very quickly when a computer is about to fail, the server C: drive is almost full, or a PC is consuming ten times the normal bandwidth. With good monitoring comes good I.T. support. Wouldn't you rather have your I.T. service call and tell **you** that they've found a problem on your server, and they are fixing it right now? Or would you prefer to have the server crash first, and then try to get a hold of someone to come fix it? Again, it's the cost of prevention versus the cost of cleanup.

Most monitoring software also allows for remote support; that means you can have an I.T. firm resolve an issue remotely without ever making an on-site service call. Almost every I.T. service firm charges less for remote support than they do for on-site support.

Even if you have an I.T. person on staff, you can make him or her so much more efficient and effective by having your computers monitored by a monitoring service.

This is such an easy decision—a few dollars per month for monitoring that will prevent a major server crash disaster.

And when your people are more efficient with time and resources, you save money.

### Does Your I.T. Firm Know What To Back Up?

I've found many misconceptions about backups. The basic concept of a backup is pretty simple—you make a copy of your data in case you accidentally delete something and need to get it back.

But the devil is in the details. What do you back up? How often? How many sets of backups should you keep? How much space do the backups need? Where should the backups be stored? How easy should it be to restore from backups? Who keeps track of the successful or failed backups? What do you do when you run out of space for backups? Are you *actually* backing up *all* of the business's critical data? So, let me provide a bit of clarity for doing backups. More backups and more redundancy is usually better, up to a point. Your backup system needs to be able to make backups that cover several months, if not years.

I often hear people say "Sure I do backups; I copy everything to Dropbox." Repeat after me: "Dropbox is not a backup system." If you make one copy of an important file and put it on Dropbox, that's all you have—one copy of your document.

A backup system will duplicate important files, and then make them generally inaccessible to you (so that you can't accidently modify or delete your backup). It will make multiple dated copies of those backups, so that you can go back six months or a year to find a previous version of an important file. Dropbox and similar cloud storage systems don't do any of that.

Additionally, having some backups on site and some backups stored off site is critical. Backups stored on site are convenient, whether your backup destination is tapes, network storage, or external hard drives. The off-site backups could be a set of hard drives stored in a vault, a dedicated cloud-based backup service, or an off-site network-attached storage device. Off-site backups are for major disaster recovery, such as fire, natural disasters and the like.

Backing up your shared files, images, X-rays, patient records and accounting software is a bare minimum. Most I.T. firms that are performing backups do that much. However, most Small Businesses haven't given a thought to backing up email.

Way back before electronic communication, many Small Businesses kept copies of every letter going in and out of the organization. That is, they kept copies of correspondence in a file cabinet; every year, they would take a load of old folders and put them in archive boxes to be stored offsite.

Somehow, when email became our main form of communication, that practice was lost. Now, with most companies using Outlook, old emails get deleted, copied to an archive file and then forgotten about, or just lost. You might assume that because you are doing backups on your server that your email is being backed up, too. However, even if that's true, it's likely that the whole email server database is being backed up, which makes restoring one missing email a huge ordeal.

The best email backup solution is a dedicated email backup device or service.

One of my clients asked me if I had access to emails sent five years ago. Since we were using the Mailstore email archiver, I told him we definitely had those emails. I asked him for some search criteria—sender and recipient email addresses, dates and a few keywords. We came up with about 10 emails that fit his criteria. Excitedly, he pointed and said "That one! That one right there!" I opened up the email, and he read it and then yelled "Yessss!!! That's 50 grand!" My client was in a dispute with a vendor over who had to pay for an installation, and the email proved that the vendor was responsible. The mail archiver (with its speedy search feature) saved my client $50,000.

So, what else needs to be backed up? Is your server's Operating System being backed up too? That means if the server crashes, can it be restored to the last backup in a matter of minutes. All modern backup systems have this capability. And don't forget to store a copy of the OS backup off site.

How about configurations for the router, switches and phone system? Are those being backed up too? Put it on your list.

Lastly, some firms go so far as to back up every workstation in the company. If you have the storage space for this, then do it. Backing up every workstation brings you very close to zero downtime.


## Does Your I.T. Firm Check Your Backups Religiously?

So, your I.T. firm is doing backups. Of course they are. Everyone says they do backups. But no one really checks them. The "set it and forget it" promise only works if someone is watching the backup results. Who is watching yours?

This is the one thing your I.T. firm can do that can make the biggest improvement in your technology situation for the least amount of money and hassle. Have them monitor your backups.

Most modern backup programs will send you an email on success or failure. Make sure your I.T. firm sets that up and tests it. Send the backup results to yourself or the I.T. administrator—better yet, both.

Top-of-the-line backup programs are *monitored* on a console. If your I.T. firm doesn't have a backup program like this, make sure they get one. For the clients we work with, every backup job is monitored in the console; success shows up as green, failure is red and warnings are yellow. In a few seconds I can see the results of hundreds of backup jobs.

One other thing to think about—are your backups

kept on site or off site? If you don't have a backup copy off site, you have no way to recover from a disaster. Without going into all the details about this, make sure your backups are being sent to cloud storage or that you have a recent set stored off site.

Again, knowing *what* to back up is critical. Imagine spending thousands of dollars on the right backup program, data storage, and I.T. expertise to set it up. But the day comes to recover a corrupted accounting file, and your less-than-attentive I.T. firm tells you, "Well, we *thought* we were backing up the accounting data, but it's not there. Last summer, when we moved the accounting to the new server, we forgot to change the backups." Poof. Your accounting data is gone. Recovery from this is painful—hours of looking through reports, re-entering data, calling banks, digging through boxes of invoices. All because backups were not checked and verified.

A backup check consists of doing a restore of a critical business file to confirm the backups are actually working. Scheduled backup checks are a simple thing to do, but only firms that practice Wellness I.T. actually do it.

It should be obvious how critical good backups are, right? Your company *is* your data. You lose that, and you're out of business, or have to pay thousands to try to recover lost data.

### Does Your I.T. Firm Recommend Better Hardware?

Buying more expensive computers to save money sounds totally counter-intuitive, doesn't it? Well, the truth is (and I can say this with the confidence of 15 years of experience) is that cheap computers **cost you more.**

They cost more in two ways. First, cheap PCs fail

more quickly than quality PCs. I have a storage room full of dead cheap laptops that failed after two years of use. The reason a laptop is cheap is because the components are cheap, and they **will fail** more quickly.

The other way cheap hardware costs you more is that it becomes obsolete more quickly. A typical low-end laptop will be useful for 2 or 3 years, and by the end, it will feel under-powered and really slow. Slower and more problematic hardware also costs more because it requires more I.T. time to deal with. Not only will you as the owner of the computer be less productive, you will also require more I.T. help to keep that computer running.

In the end, cheap computer hardware isn't worth it for a business. Even if you are on a tight budget, you will find that money for computer hardware is better spent on middle-of-the road or higher-end computers and other technology hardware.

### Does Your I.T. Firm Know *Which* Services to Move to the Cloud?

Should you move everything to the cloud? Should you keep it all in-house? What *should* you do with cloud services? It's hard to know without getting advice from an expert.

Some things are really easy to move to the cloud, and other things, not so much. Moving email to a cloud service makes a lot of sense, and has immediate financial payoff. Just add the cost of a mail server, the cost of maintaining the mail server software, the cost of a spam filter and annual updates, plus the cost of downtime when things aren't working quite right, and compare that with the monthly cost of hosted email. Probably an easy decision.

On the other hand, if you do large digital X-rays or work with huge data files, that's not nearly as convenient to store in the cloud. Even with a fast internet connection, file transfers take a **long** time. So that kind of production needs in-house storage, which also means local backups.

Deciding what to put in the cloud and what not to takes some forethought. If you move the right services to the cloud, you can save money on those services. (And, if you move the *wrong* services to the cloud, it will cost you.)

**IS CLOUD COMPUTING FOR YOU?** Is your business ready for Cloud Computing... or is it something which could actually slow your business down? Find out with this FREE "Is Cloud Computing good for you?" checklist. Download here. www.nonlineartek.com/is-cloud-computing-for-me

## Does Your I.T. Firm Have Established Documented Procedures?

One of the most costly situations in a company is having re-do work that was done incorrectly the first time. Most mistakes like that can be avoided. How? If your I.T. firm has written documented procedures, they will be less likely to make these kinds of mistakes. It sounds really dry and boring, but having written procedures, such as what to do when a new user comes on board, how they test the backups, and what to do when an employee leaves can literally save **you** thousands of dollars.

If your I.T. firm has written procedures and checklists, then that will ensure that each time the work is done, it's done the same way *and* each person who does the work does it the same way.

How many times have you called for tech support and gotten something half-fixed? So, you call back to the I.T. person (whether it's in-house, or an I.T. service company), put in another support request, and incur the cost of having the work done again, the right way this time. You're paying twice to get the work done.

Ask your I.T. firm if they have documented written procedures. Or, better yet, bring on an I.T. service who already has documented procedures in place, so you know things will be done right the first time.

## Does Your I.T. Firm Work With You to Create a Future Plan?

Do you know which technologies to invest in, and which ones are losing ground? Between two competing ideas, which one will win? Do you have time to stay on top of all of this?

If you are not getting technology advice from your I.T. person on a quarterly basis (at least!), considering bringing on a third-party technology advisor.

The advisor can work with your existing I.T. teams and services to provide direction on where things are headed, and when and where to invest in new technology.

As I mentioned previously, many I.T. firms act like the plumber – they come in, fix a problem, and leave, without ever having a conversation with you. The best I.T. firms act as the Trusted Advisor, similar to your accountant, to help you make the right decisions for your organization's future. Always choose the I.T. "Trusted Advisor" over the I.T. "plumber."

### Does Your I.T Firm Have Complete Transparency on Issues and Billing?

When your I.T. firm works on something, do you know the status of that project or task? How easy is it to find out? When you receive your I.T. service bill, do you recognize every issue on the bill? How easy is it to find the details of what's on your bill? Is it a quick online lookup, or does it take five emails, two phone calls and three weeks of back-and-forth to find out?

If your I.T. firm is using a proper issue tracking system (a "ticket system"), you should have complete access to your tasks and be able to know the status of any issue with just a few clicks in the online portal. That's just good service.

I'm always surprised how many I.T. firms are *not* using a proper issue tracking system, and rely on random emails, text messages and post-it notes to keep track of important client tasks and projects.

### Where To Go From Here

How does your I.T. firm measure up? If they have a perfect score, then you have a top notch I.T. services firm. Congratulations! Consider yourself fortunate, and enjoy the benefits of good security, and a smoothly running infrastructure.

If you're missing a few points covered in the checklist below, don't worry. Many I.T. firms will rise to the challenge, especially if you are armed with a checklist of how they can improve.

If your I.T. firm is missing most of the checklist points, it's probably time to start the search for an I.T. firm that has your best interests in mind.

Download the checklist that my company uses to provide advice and support on best I.T. practices.

**IS YOUR FIRM FOLLOWING BEST PRACTICES?:**
Compare your I.T. firm's work with this continually updated I.T. Best Practices Checklist.
www.nonlineartek.com/best-practices-checklist

# Five I.T. Practitioners—Which One Is Right For You?

While I can't claim to be the logical choice for every Small Business, I do know that I.T. Wellness support works best in the long term, and will give your business the best results.

As long as I.T. firms subscribe to the principles of I.T. Wellness, they will serve you well. See which of the I.T. firms below match best with your kind of business.

## The Big Local I.T. Firms

Bigger I.T. firms have more resources available to them, and are able to be more places at once. That's the advantage of these firms. Big firms also have pretty good support, and generally know how to handle big projects.

However, they have some challenges too. For example, a large I.T. firm will have technicians whose abilities vary widely. Depending who shows up to help you, you might get an expert or you might get a beginner. The larger I.T. firms tend to do technology well, but "HR" is their week point—they struggle to get consistently good technicians.

That affects your business because you may see inconsistent technical support—sometimes the problem is solved quickly by a smart polite techie, and sometimes you will see the problem drag on (or even made worse) by a beginner tech.

## The Small Local I.T. Firms

The small local firm can be anything from a "one-man shop" to a small company of a couple technicians to a company of 10 or 15 employees. The ad-vantage of the small firm is the personal support and deep knowledge these firms have about the systems they support. The downside is that a smaller firm cannot be everywhere at the same time, so to support a very large client can be more challenging.

## The College Student

When Small Businesses first start out, they often employ a college-age relative or friend of the family to do the I.T. work. The advantage of this kind of support is that they are usually very cheap. However, they typically do not use I.T. Wellness principles because the arrangement is "call-as-needed" from the start. Additionally, their experience with older systems is often lacking, and they don't wish to dig into older systems to learn how to support them.

Also, they have a high probability of leaving for a permanent job elsewhere as they finish up school and get "a real job".

The college student will likely use quick-fix solutions and not take long-term return on investment into account. For example, they might purchase a cheap PC instead of a more expensive one that has a US-based customer support infrastructure in place. When the cheap PC fails, the time (and thus money) required repair it far exceeds the savings of buying the cheaper PC.

## The Employee Who Should Be Doing His Own Job

Many small companies have a designated "tech support" employee—someone who knows how to troubleshoot network and computer problems, and can fix most of them.

The trouble with this approach is that that employee probably has more important things he should be doing with his time. That is, if he is a dental assistant, he should be working on patients. If he's in marketing, he should be doing his marketing work. That employee will likely do the quickest thing he can do to resolve the problem and get back to doing his real job.

Most of the I.T. disasters I have had to clean up were due to quick-fix solutions in places where I.T. Wellness principles were not implemented. Quick fix solutions were often to blame for these disasters.

One of the best ways we can work with Small Businesses is for my firm to handle the proactive support parts of the job, and still allow the designated employee to be the front line of support. That means that employee can handle printer installations, new email account setup and a simple software install, but they're not responsible for setting up a backup system or network security. Additionally, if an outside I.T. firm like mine is familiar with your business, we can help out immediately in the case of an urgent technical problem.

### The Boutique I.T. Firm
The boutique I.T. firm specializes in a small number of kinds of clients; they become familiar with the concerns and working styles of their clients, and get to know their clients' specialized software well. If you own a dental office or chiropractic office, you may be best served by a boutique I.T. firm that specializes in your kind of work.

### Which Kind of I.T. Firm is Best For You?
Clearly, some kinds of I.T. firms are better than others. Having no I.T. firm to help you, or assigning I.T. roles to a reluctant employee or family member

are not good choices for moving your business forward.

Large I.T. firms and smaller boutique firms have pros and cons. I'm partial toward the small and boutique I.T. firms because I think we strike the best balance of technical know-how, automation and personal service. My customers think so too.

# ABOUT THE AUTHOR

John Verbrugge is an I.T. consultant in Grand Rapids, Michigan. His firm, Nonlinear Tech, has participated in the growth and technical needs of dozens of businesses in the West Michigan area.
John consults on technical and business growth topics, but his passion lies with helping businesses in the health and wellness industry make good technical decisions and make the best use of their technology.

Nonlinear Tech Inc
http://www.nonlineartek.com
info@nonlineartek.com
616-528-1566

Made in the USA
San Bernardino, CA
19 August 2016